Periodic Table

TEACHING CHEMISTRY IN A DIVERSIFIED CLASSROOM

BOOK 3

By Sammie Jacobs

Copyright

All rights reserved under International and Pan-American Copyright Conventions. By payment of the required fees, you have been granted the non-exclusive, non-transferable right to use the materials provided for the preparation of lessons and during the direct instruction of students.

PERIODIC TABLE. Copyright 2020 ©

Student worksheets and assessments may be duplicated for classroom use, the number not to exceed the number of students in each class. Notice of copyright must appear on all copies as provided on the page.

Activity kits duplicated and assembled as directed should display the copyright markings as prescribed in the directions and on the materials.

No other part of this text may be reproduced, up-loaded, displayed, shared, transmitted, down-loaded, decompiled, reverse engineered, or stored in or introduced into any information storage and retrieval system, in any form or by any means, whether electronic or mechanical, now known or hereinafter invented, without the express written permission of Lavish Publishing, LLC.

First Edition
2020 Lavish Publishing, LLC
Teaching Chemistry in a Diversified Classroom book 3
All Rights Reserved
Published in the United States by Lavish Publishing, LLC, Midland, Texas
Select Illustrations by Sanghamitra Dasgupta
Cover Design by: Victor R. Sosa
Cover Images: CanStock
Paperback edition
ISBN: 978-1-64900-002-6
www.LavishPublishing.com

FOREWORD

Teachers, parents, and educators of all kinds – welcome to my classroom!

Here within these pages, and in fact this entire series, are lessons and activities that I have created and used with my own students over the years presented in an easy-to-use format. Through trial and error, hardship and success, I have learned how to present the difficult world of introductory chemistry to students in scalable terms, ways that lend themselves to a wide variety of learners.

So, if you teach in a small class, large class, average class, have ESL or SPED, or even if you are homeschooling, you will find the flexibility of my lesson plans tailor made for you. Each unit includes a complete calendar plan, lessons for each classroom day for forty-five minutes of instruction each, and even formative and summative assessments to check for learning. Simply use the lessons, scale them as need be, and augment whenever you like.

Be sure to join my Teaching Chemistry PLC on Facebook for even more ideas and sharing, and of course, I hope for the best with you and your students!

Sammie Jacobs

TABLE OF CONTENTS

Unit Tools . . . . Pages 1 – 5
Calendar Key and Usage
Guide Calendar
Blank Calendar
Periodic Table Vocabulary

Lesson Plans . . . . Pages 6 – 45
Mendeleev and Moseley
Periodic Table Tricks
Color and Label Big Periodic Table
Periodic Trends
Magic Board Practice with Answers
Periodic Table Review & Exam

Assessments . . . . Pages 45 – 50
Mendeleev and Moseley Quiz
Periodic Table Family Quiz
Periodic Table Sections Quiz
Periodic Table Tricks Quiz
Periodic Table Trends Quiz
Periodic Table Vocabulary Quiz
Periodic Table Unit Exam
Assessment Answers

Unit Calendar

The **Unit Calendar** is an important organizational tool.

It features a list of vocabulary terms (top section) for the unit, the plan of days (center section), and the student learning goals (bottom section).

A **completed teacher copy** of the calendar with the unit overview is provided. On this version, days are not written as dates, but as lesson days. However, on my copy that I use in my classroom, these have been laid to the school or district calendar and those day numbers are replaced with actual meeting calendar dates. There are five in a row, which is a work week, and days that we do not meet or have instruction time are crossed off.

You will notice that the unit calendars are always a whole number of weeks long and every day has a lesson planned, but often you will have holidays or non-class days. When you have less days, combine lessons where appropriate, or omit lessons that are not included on the student goals and exam if necessary. We also only meet for forty-five minutes per day, so the lessons are designed to fit within that time frame. If you have longer class periods, you may decide to augment the days with additional activities, such as vocabulary games.

The **blank student copy** can be used to customize the calendar to match your available days. It can also be copied and distributed to students, along with the vocabulary lists with definitions.

Calendar ideas:
1. Have students fill in the calendar daily with daily topic as they begin class as a way of making them aware of what the day's lesson will entail.
2. Have students fill out the calendar at the start of a new unit so they have the plan ahead of time and can be proactive with their learning.
3. Give students a completed copy with the appropriate dates and topics at the start of a new unit to save time.
4. Have students highlight each day and the corresponding terms on the vocab list with a different color (color coordinating them) for reviewing later.
5. Have students use the calendar to record assignments for each day and highlight them as they are completed and submitted.
6. Have the students write a question for each day to reflect upon later.

Unit **Periodic Table**   Month

Vocabulary

alkali metals	halogens	Moseley
alkaline earth metals	inference	noble gases
atomic number	ionic radius	non-metals
atomic radius	ionization energy	nuclear symbol
chemical family	mass number	periods
electron affinity	Mendeleev	transition metals
electronegativity	metalloids	trends
groups	metals	valence

Daily Agenda

M 1	T 2	W 3	Th 4	F 5
Mendeleev And Moseley How and Why the Periods Come Together	Table Tricks Valence e- Charges Families Metals and Non-Metals	Color and Label the Big Periodic Table Valence e- Charges Families Metals and Non-Metals	Periodic Trends Ionization Energy Atomic Radius Electronegativity	Periodic Table Review And Exam

Learning Goals

Target Concept	none	weak	solid
Explain how Mendeleev organized it and what it did			
Explain how Moseley organized it and how it improved it			
Label and describe the three properties of the 3 main sections			
Identify the 4 families and transition metals			
Identify, explain, and use periodic trends			

© Lavish Publishing, LLC

Unit __PERIODIC TABLE__   Month _____

Vocabulary

alkali metals	halogens	Moseley
alkaline earth metals	inference	noble gases
atomic number	ionic radius	non-metals
atomic radius	ionization energy	nuclear symbol
chemical family	mass number	periods
electron affinity	Mendeleev	transition metals
electronegativity	metalloids	trends
groups	metals	valence

Daily Agenda

M	T	W	Th	F

Learning Goals

Target Concept	none	weak	solid
Explain how Mendeleev organized it and what it did			
Explain how Moseley organized it and how it improved it			
Label and describe the three properties of the 3 main sections			
Identify the 4 families and transition metals			
Identify, explain, and use periodic trends			

© Lavish Publishing, LLC

Periodic Table Vocabulary

word   definition

alkali metals — Group I of the Periodic Table, the ____, is composed of highly reactive metals.

alkaline earth metals — The ___, the 2nd group on the periodic table, are not found freely in nature.

atomic number — ____ is equal to the number of protons in an atom; used to arrange the periodic table.

atomic radius — The ___ of an element is half of the distance between the centers of two atoms of that element that are just touching each other.

chemical family — Elements in the same group or column on the periodic table are said to belong to the same ___.

electron affinity — ___ is the quantitative measure, usually given in electron-volts, of the tendency of an atom or molecule to capture an electron and to form a negative ion.

electronegativity — ____ is the tendency for an atom to attract electrons to itself when it is chemically combined with another element.

groups — Columns on the periodic table are called ____.

halogens — The elements of Group VII of the Periodic Table are called ___, which means "salt formers".

inference — ___ is the process of deriving the strict logical consequences of assumed premises (what do you think happened / will happen based on the data).

ionic radius — The ___ is the size of the radius of an ion.

ionization energy — The ___ is the energy required to completely remove an electron from an atom or ion.

mass number — ___ is equal to protons plus neutrons; it is the average atomic mass rounded to a whole number.

Mendeleev — A Russian born scientist named Dmitri ___ created the first periodic table of elements based upon the atomic mass of the elements - he was able to use this to make predictions about the behavior of elements.

metalloids — ___ are found along the stairs (right above or below) on the periodic table - they take on a variety of characteristics from both metals and non-metals.

metals	___ are found UNDER the stairs on the periodic table - they are shiny, hard, have a + charge in ion form (givers), and are good conductors of heat and electricity.
Moseley	Henry ___ was able to take Mendeleev's periodic table and rearrange it in order by atomic number - which made it work better - and is still used today.
noble gases	The 8th group or family are known as the ___ - because they satisfy the 'octet rule' alone, they do not "hob-knob" with other elements or form bonds.
non-metals	___ are found on top of the stairs on the periodic table - they are dull, brittle solids, can be gases, have a negative charge in ion form (takers) and are poor conductors of heat and electricity (insulators).
nuclear symbol	The letter or letters that represent an element are called the _____, which can include details about the structure or parts of that atom.
periods	The rows on the table of elements are called _____.
transition metals	The ___ are found in the center of the periodic table and get their name because they have a tendency to change their oxidation numbers by moving their valence electrons around.
trends	On the periodic table, we can see that there are several ____ that repeat, period after period and pointing either at Friendly Fluorine or Fatty Francium.
valence	The outer energy level of an atom is called the ____ shell and the electrons that are in it are called ____ electrons. (same word for both blanks).

© Lavish Publishing, LLC

Key to Lesson Plans

Topic: coordinated to the unit calendar - what we are learning about today?
Day: coordinated to the calendar - sequence of presentation out of total available
Unit: coordinated to the unit calendar - unit the lesson falls under

Learning Target:
A postable **learning objective** that gives an outline of what the day's lesson will cover and **artifact** the students will produce for evaluation. These can be used as is or modified to suit your student output or to focus on a different aspect of the lesson, as these often only cover one part of a layered lesson and group of activities.

Student Goals:
At the bottom of the calendar, the students have a list of **goals** they want to meet by the end of the unit, which are covered on the **unit exam**. This section coordinates which goals are being addressed during this lesson. This section also lists the **vocabulary terms** from the calendar that will be defined or needed during the lesson so they can be connected, pre-taught, or directed to study afterwards.

Agenda:
A postable list of what activities this day will include. They are divided into two main types of instruction:

Lecture **(L)**, where the teacher is providing direct instruction and the students are actively listening, taking notes and providing feedback at given intervals.

Activities **(A)**, where the student is producing work of some kind and the teacher is observing and providing support when needed.

Student Materials:
A list of materials that each student, pair, or group, will need to complete the lesson. They will need to be prepared ahead of time and be ready to **pick up** as students enter the room or to be **distributed** at the appropriate time during the lesson.

Generally, if it is a **one-to-one item**, I have them ready and students pick them up as they come in to save time. **Paired and Group materials** can be handed out or placed in a location to have a student go and get while the teacher is preparing some other part or completing a task during transition.

Props:
These are items you generally only need one of and can be hidden to pull out at the appropriate time or placed on a table that the students can visit before and after the lesson.

They are support items that deepen the understanding of the lecture and either generate questions or provide answers.

Having props is vital for a wide range of learners, so do not feel limited to what I have and use – explore and add anything that you want to aid in this process, as all students benefit. Save your props as you build them, as they often are used in future lessons to bring concepts forward and tie ideas and understanding together.

I have a set of props that stay on my front table throughout the year, as I pick them up and refer to them often: a plain bottle of water that is labeled H_2O, a bottle that is half cooking oil and half water, a bottle that is water with about a gram of dirt in it, and a bottle that is water with about a gram of corn starch in it. Other props are added and removed as needed.

Actions and Rationale:
These are the key points of the lesson and the reasoning behind what is being said or done. All activities will have a separate directions page if needed, as well as a reproduceable student copy and directions for building all props and student materials.

During each unit, we use a variety of learning and presentation styles that include having the students listen, speak, read and write. There are also a variety of study and practice skills woven into the lessons that students can learn to use in and outside of their Chemistry class.

Topic: Mendeleev and Moseley
Day: 1
Unit: Periodic Table

Learning Target:
I can understand and explain how Mendeleev and Moseley contributed to the creation and organization of the periodic table, which give it predictive powers.

Student Goals:
Students will need to know who Mendeleev and Moseley are, how they organized the periodic table, and what strengths / weaknesses each form possessed.

Agenda:
A – Warmup: Unit Calendar
A – M & M Reading & Research
A – M & M Discussion & Recap

Student Materials:
Journals for notes
Articles or books for reading
Phones or computers with internet for research if desired

Props:
T-chart of M & M for discussion

Actions and Rationale:
Warmup – unit calendar. Since this is the start of a new unit, we fill in the calendar and I make announcements about the week and exam. Some years, the first semester is too short, and I have to break this unit up rather than teach it as a block – a day here and a day there, with the material added to the exam for the unit I taught it in. However, it is good to have this as a separate small unit if possible since it chunks the material into a nice block of related information.

This week, we will be learning more classroom procedures, preparing more tools, and trying out various forms of instruction that are not direct lecture. Lectures get boring for them and you, so this week we will be looking at reading to gain information and take notes, so we transition into directions for today's activity by talking about how science is an ongoing process.

I tell them that there are usually many who work on any given problem and that were maybe dozens who looked at organizing the periodic table in one way or another, but we are going to focus on two of them (Mendeleev and Moseley) who I call M & M – M&M Created the Periodic Table.

Activity - reading & research. Explain how the research will take place and the expected outcomes. I have an article in my textbook that we use for this, but you can locate one or more online, or have the students search for it themselves. Or you could turn this into a video lesson if you wanted to hunt down one or more of those. I have them put this information in their journal on a T-chart – they can put anything else that they want, as you will go over it in the discussion and make sure they have at least these main points:

Their goals of focus are –
Who are M & M?
How did they organize the Periodic Table?
How effective is their version of the periodic table for predictive powers?

Mendeleev	Moseley
Russian	English
Worked on PT 1st	Adapted the PT 2nd
Used Ave Atomic Mass to organize	Used Atomic Number (p+) to organize
Had some predictive powers, but it had problems that he had to 'fix' by moving a few elements	Has strong predictive powers, but had blank spaces for elements not then discovered
no longer in use	still in use today

Discussion thoughts:

Mendeleev made the PT, Moseley 'moseyed' in and made it better

Both left gaps for undiscovered elements, but Moseley didn't have to fudge anything – it was perfect

Moseley's predictive powers were so good, we still use it today.

Activity – discussion. You can have the students compare to each other and expand or fill in what they missed, then have a large group meeting to make sure everyone got the details above, which will be covered on the exam. Expand or shrink this part however you like, but I always make sure I go over those key points.

If you have time and want to practice further, there is a magic board activity (worksheet) located with day 4's activities.

Topic: PT Notes & Table Tricks
Day: 2
Unit: Periodic Table

Learning Target:
I can label and describe many features of the periodic table, including sections, families, predicting charges, and valence electrons.

Student Goals:
Students need to know the families, sections, charges and valence electrons for the periodic table, as well as main groups and properties for each.

Agenda:
A – Warmup: Mendeleev & Moseley Quiz
L – PT Notes & Table Tricks
A – Color and Label Notes

Student Materials:
Journals & glue
PT notes pages for coloring and gluing
Color pencils
1 sticky note per student

Props:
Completed PT notes page
Picture of an Atom (Bohr Model)

Actions and Rationale:
Activity – M&M Quiz. These little quizzes give me a chance to gather a little data on how our learning is going, as well as a chance to reteach the topic. The students fill out the quiz, I take them up, then we go over the answers together. I grade them later for a practice grade, score 1, 2 or 3. After you have gone over them, transition to the lecture, which is those 'predictive powers' we talked about.

Lecture – PT notes and table tricks. You will want to spend a few minutes on each section telling them about the four small periodic tables and what that information means, but don't get TOO detailed yet. We spend weeks learning this material, so this is only a small introduction for the concepts of Valence Electrons and Ion Charges. For now, they just need to know how to discover those using the periodic table by using **table tricks** (counting).

Table 1 is for **families**, which are groups of elements that have similar chemical properties. This also serves as a practice run for coloring our big periodic table tomorrow, so have them color it EXACTLY as I have shown (ROYGBV & white). This matching color scheme will help later when you are using the periodic table. For now, give the names and color scheme, but we will come back to this later.

Table 2 is for **sections**, which are metals, non-metals, and metalloids. I also have them put the properties for each on a sticky note that goes with this one. Notice that there are families that are all metals, the noble gases are all non-metals, and the 'no names' are broken by the stairs that divide the PT into these 3 sections.

Metal Properties – shiny, hard (solid), malleable, conductors of heat & electricity
Non-Metal Properties – dull, gases, brittle solids, insulators
Metalloid Properties – falling along the stairs on either side, they have properties of both and are **hard to predict** because they borrow from either side

Table 3 is for **valence electrons**, which we are focusing on the foundation only today – counting. Tall Columns (A or representative elements) follow strict rules with their electrons. They should know what electrons are based on previous years but having a Bohr Model type picture of an atom to refer to will help to jar their memory, so you can project or draw mine for them.

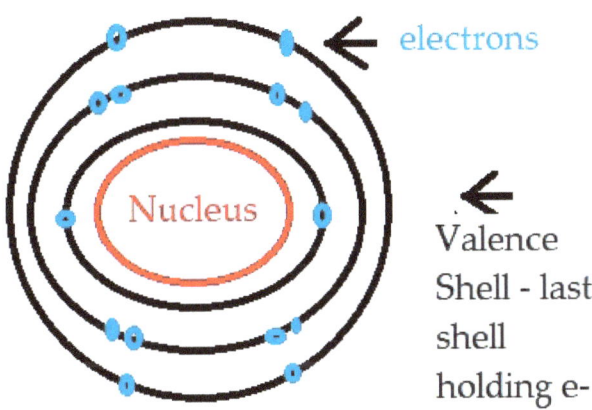

The **Valence Shell** is the outside or last shell, and that is all we are doing – counting how many each element in the tall columns has in theirs. Tell them as much or as little about this as you like, but I find keeping it simple at this point is best – all we are doing is counting the tall columns and that is called the **number of valence electrons**.

Another activity that I typically do with the **electrons in atoms unit** could be added here if you have more time. I give out tootsie-roll pops and we eat them, talking about how they are similar to an atom – the tootsie-roll center is the nucleus, and the outside that we are eating is the valence shell, with the candy coating like layers and layers of electron shells (orbitals) between them. If you don't have time now, keep it in mind to come back to later, as the students enjoy it and it does help solidify their understanding of the atomic model as a whole.

Table 4 is for **charge**, which comes from those valence electrons, but they are NOT the same thing. Again, we will be doing this for weeks (months), so don't worry about the deep details. For now, they just need to know how to use the periodic table to get the charge, which are marked here. The story I tell them is worth it if you have time, as it can really help give them a concrete version of this concept.

The Charge Story – I like to do this as a discussion, so I ask the questions and guide their responses where I want to go.

How many of you go to lunch with your friends? (yes, they do lunch with friends)
When you go, who pays? (usually each buys their own)
What if your friend buys your lunch for you? That would make you feel good, right? That's because your friend would be a giver - they give you lunch – and **givers are positive people**.

What if your friend took your lunch or expected you to buy theirs? Would that make you feel good? Probably not, because that friend would be a taker, and **takers are negative people**.

And so it is with atoms and valence electrons, which form what we call 'ions' – ions are either givers or takers. If they are givers, they are positively charged, and if they are takers, they are negatively charged (we sometimes call these 'the charges'). But which are what on the periodic table? Did you know you can tell just by where they are located? That is part of the predictive power the periodic table has.

If they only have a few valence electrons, they are going to give them away. Point to the periodic table with the valence electrons as you explain this and have them look at theirs. Which only have a few valence electrons? That's right – the metals end. They will all give their valence electrons away, which gives them a positive charge. Metals are givers.

If they have a lot (more than 4), they will take or steal a few more to get to 8. Which ones are going to be those takers? That's right – the non-metals above the stairs are going to have negative charges. Non-metals are takers.

Why does this work? We have what we call the **octet rule** – it says that all atoms want 8 valence electrons in their valence shell to be like the noble gases. Those non-metals will steal enough to fill up to 8, and those with only a few will ditch them and drop back to a nice full shell. Point to your Bohr model picture as you explain this.

Finally, point at the 4 column. This is the tricky one, because they are in the middle and can't decide what to do. They will never form ions because of this. Below 4 are positive (**cations**, remember by 'cats have paws' so they are **paws**-ative) and above 4 are negative (**anions**).

Fun Extras – these are things you will want to share with your students when you have time, when the moment is right, either now or during later lessons. They are good things to know and interesting ideas or facts that can deepen understanding.

Why do the **Noble Gases** not have a charge? They don't give, take or share, which means they don't form bonds, so, they have no charge. They are like the queen, not hanging out or 'hobnobbing' with the commoners. Interesting note – noble gasses are actually only commonly found naturally in earth's atmosphere, and you **can** make some of the bottom of the family form bonds, but they don't do it willingly; it takes effort or intervention.

Why does the counting trick not work for **transition metals**? They are named 'transition' because they move their electrons around and by not following the rules. We are going to learn lots about them in a few weeks, and they will always be tricky!

What about **Hydrogen**? He is the top of the Alkali Metals family, but I have colored him a different color – why do you think that is? Hydrogen is a double agent. He plays on both sides – sometimes as a metal, and sometimes as a non-metals. I call him the 'wild child' because he is an exception to lots of rules. Let's make sure we are watching out for Hydrogen as well and talking about this oddity at some point.

Broken families with those elements below the stairs are actually metals and will form cations, so the counting trick for charges only applies to those non-metals above the stairs. You can try to teach this now, but I skip it and we focus in on it more later when we learn about transition metals.

Again, if you have time and want them to practice, there are work sheets (magic boards) available at the end of day 4.

Periodic Table Notes

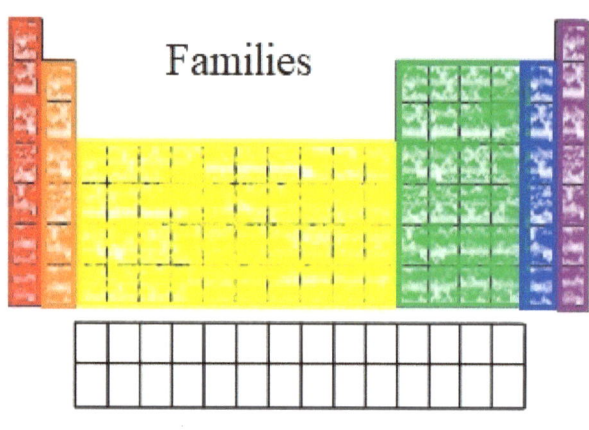
Families

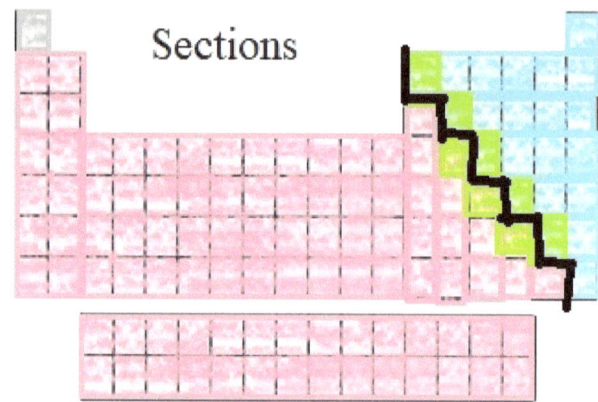
Sections

- ● Alkali Metals _____
- ● Alkaline Earth Metals _____
- ● Transition Metals _____
- ● "No Name" Families _____
- ● Halogens _____
- ● Noble Gases _____
- ○ Inner Transition Metals _____

- ● Metals _____
- ○ Non-Metals _____
- ● Metalloids _____

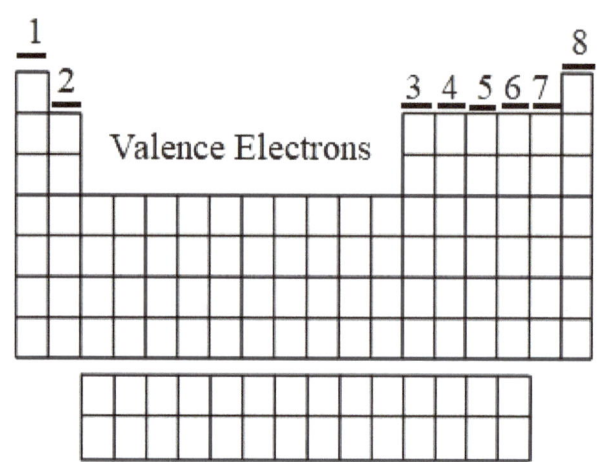
Valence Electrons

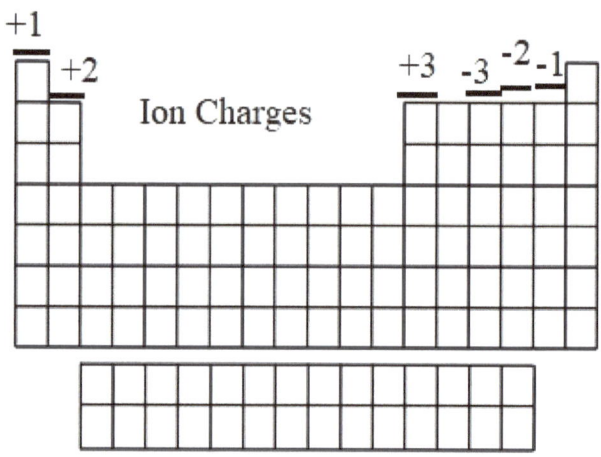
Ion Charges

14

Properties of Elements

Metals	**Non-Metals**
shiny	dull
hard (solid)	soft (solid) or gas
malleable	brittle
conductors of heat & electricity	insulators

Metalloids

Fall on the line between and borrow from both, so HARD TO PREDICT

Sticky note for students to put onto notes >>>

Periodic Table Notes

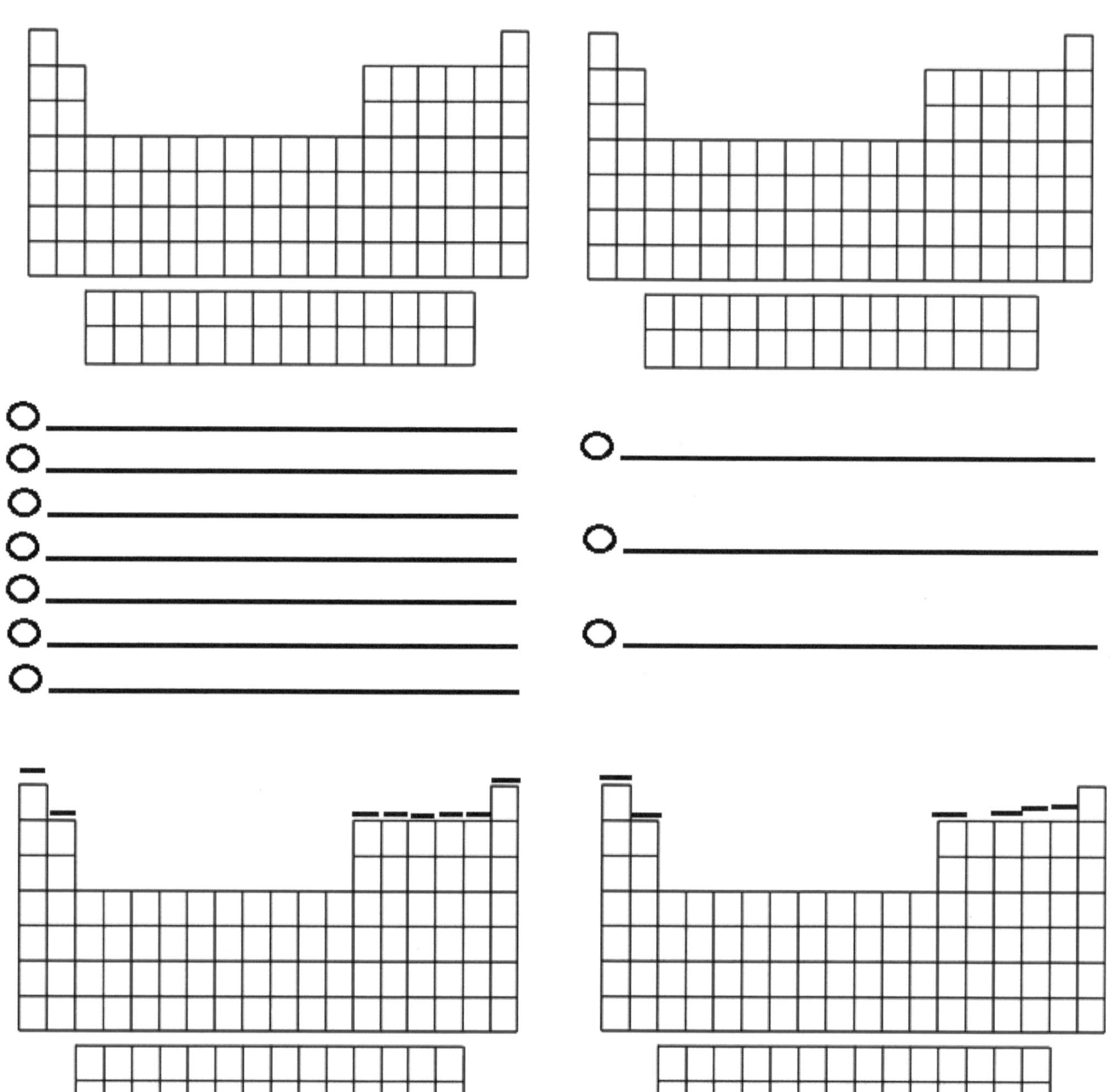

Topic: Big Periodic Table
Day: 3
Unit: Periodic Table

Learning Target:
I can color and label my reference periodic table to identify the families, valence electrons and ion charges.

Student Goals:
Students need to know the names of the families, how to count valence electrons, and what charge each family is going to have for the exam, and more importantly for the rest of the year to use (things get a lot harder for them if they can't do this).

Agenda:
A – Warmup: Practice Quiz
L/A – Color & Label Big PT

Student Materials:
Journal with reference envelope & big periodic table*
Color pencils (I recommend against markers)
Two colors of highlighters

* A **Note** about periodic tables – you can use any plain white one that you like, but I recommend keeping it simple. The boxes need to have the Atomic Number, Symbol, and Ave Atomic Mass. Anything else is just clutter. We have a 'state' version in Texas that came with our end of course exam, so if you need one or your state doesn't have one, visit the TEA STARR Reference Materials Chemistry if you would like to use it. It is free – you just download and print. It also has four parts with valuable information, and we will use them all during this series. Ours have the PT on the front and the section containing the polyatomic ions on the back.

Props:
Your Big Periodic Table perfectly colored and labeled

Actions and Rationale:
Warmup – practice quiz. You have a couple to choose from. You can do one or both for families and groups. I take them up for points before I go over them, then transition into transferring what we learned about the periodic table over to the large one we will keep in our journal for reference.

Lecture / Coloring Activity – students will color their big periodic table to match the notes. Remind them to be CAREFUL when they do this, as it needs to match exactly what it should have been. Later when you are using it, they will know which column is 'the red one'. They also need to make the key with colors and labels for the families with the pencil they used to color so they match.

I have them write the words **Valence** and **Charges** and mark them with two different colors of highlighter to code the numbers at the tops of the columns. Tell them not to worry about the sections of the periodic table – we really only care about two (metals and nonmetals above and below the stairs) and will work with them for the rest of the year. Metalloids exist, but they don't play a role in anything as a group.

Tips for Coloring:

1. Color lightly – the numbers and writing still need to be legible
2. Write the numbers at the column tops by hand (ignore any that are already there) and highlight carefully with two different colors
3. Make the color key up at the top and out of the way – we are going to add things later on the edges, so keep them clear
4. Put your name on it – that way it doesn't get lost when using them

Walk around and observe / help as they work. I give them 10 to 15 minutes (time it so they get used to deadlines). I also do a check off for practice points when they have it completed – if they don't finish in time, they have to do it and show it to me later. We finish up with a short lecture, continuing from what they learned yesterday.

First, ask, "Can you locate Potassium?" This is to get them used to finding elements that don't start with the same letter as the name – in this case K. Say, "Everyone, put your finger on Potassium." Walk through the room and see that they are participating. Ask, "What letter or **symbol** is Potassium?"

That's right, it's K. Many elements on the periodic table don't match, so get used to the names. Remember – the more you use the periodic table, the easier it gets.

Next, ask, "How many valence electrons does Potassium have?" They should slide their finger up and find valence electrons – it is 1. Then ask, "So, what's its charge?" Right, it's a giver, charge positive 1 (make sure they say the positive, not just one). Check your time – if you have enough, do more of them, some cations, some anions, having them locate and identify name, symbol, family, charge and valence electrons.

Now, lets link some of this up – tell them to look at the orange column. "What are they called?" That's right – Alkaline Earth Metals. Memory trick – the short column has the longer name. Ask, "How many valence electrons do ALL of them have?" They have 2. "What do they all do with their 2?" They give them away – all with a positive 2 charge.

Now explain - that is what gives them SIMILAR reactivity and why we call them a family. They act the same because they all have a positive 2 charge. Do you think that is true for group 1? How about the halogens? Yup – those full columns (families) get a special name because they have the same valence electrons and the same chemical properties. This is the magical prediction power Moseley gave us by lining them up using their atomic numbers.

Finally, how about all these green ones. See how the stairs cuts through them? I call them "no names" because these are BROKEN families. The stairs cut them so that the top half act the same and the bottom act the same, but they don't act like each other. That is why the ones under the stairs are all part of the metals, and the ones over the stairs are part of the non-metals, and these families are split.

Explain, "Even though these 'no name' families aren't special like the others, we might want to talk about them some time. If we do, we simply call them by the name of the element at the top. So, these are really the boron group (point to it), carbon group (point to it), nitrogen group (point to it), and oxygen group (point to it). Just in case you ever need to know that." If it is confusing, don't worry about it. We are just introducing the concept, as we will be using and building on it for the rest of the year and they will have time to clarify their understanding.

Again, if you have time and want more practice, there are practice sheets at the end of day 4.

Topic: Periodic Trends
Day: 4
Unit: Periodic Table

Learning Target:
I can identify, explain, and use periodic trends to predict how atomic size, electronegativity, and ionization energy change across the Periodic Table.

Student Goals:
Students need a beginning understanding of these three processes and how the predictive power of the periodic table can be used to compare them across elements.

Agenda:
A – Warmup: Table Tricks Quiz
L – Periodic Trends
A – Magic Board Practice

Student Materials:
Journal for Notes
Reference Periodic Table
Slips of paper for Table Tricks Quiz
Tiny Periodic tables to glue in for notes
Magic Boards for Practice Games

Props:
Periodic Table for Board or Power Point

Actions and Rationale:
Warmup – Table Tricks Quiz. They can complete while you take attendance. I take them up for practice points before I go over them, then transition into the final piece for this week, which are periodic trends.

Lecture – periodic trends. Open with, "Periodic trends are the final piece to the predictive powers of the periodic table. To learn about this, we will need to learn three new vocab words and then apply them to the periodic table, so let's start with those definitions."

Atomic Radius – this is the diameter of two identical atoms placed side by side and connected center to center. I draw this on the board, or you could make a picture to post or put up on a power point. The key here is that it means the **size of the atom**.

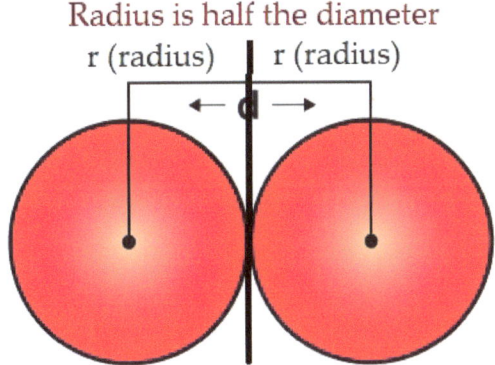

Electronegativity – this is the strength atoms use to **pull on electrons** when they share them in a bond. I use a string tied in a loop or an extra large rubber band to demonstrate this bond. We will be talking about covalent bonds and electronegativity throughout the year, and this is our first glimpse of the process.

I ask for a volunteer who comes up to stand beside me. I give them one end of the string and I keep the other, each holding it up so everyone can see with one finger put into the loop to hold it by.

I say, "Pull on your end, gently though." I let them pull and extend my arm towards them. I point to the connection and say, "This string is the bond and our fingers are the electrons. The bond is formed by sharing where we atoms each put one electron into that bond. Those electrons travel back and forth between us." I pull on my side and they extend to let me have more 'possession' of the pair. Then I let them have it, then I pull It back. Then I say, "How hard we are each pulling is called **electronegativity**." I thank my assistant and they go sit down. If I need to show it again later, I just use my two hands and shift it from side to side in front of me for the visual, or I use that rubber band since it is more manageable.

Ionization Energy – the trickiest of the three we will learn about. It's tricky because this runs the opposite of what you might think when looking at a scenario.

Basically, ionization energy is a measure of how much energy is required by another atom to remove an electron from 'me'. I always personify atoms and call myself the atom in this case. I stand in front of the class, indicating my body and say, "I am the atom and I have my valence electrons." I show my hands, waving them out beside me. "If another atom wants to take them from me, Ionization Energy describes how badly I want to keep them, or how much energy that atom will need to get it away from me."

Trends – Okay, let's turn to applying these to the periodic table. I have them glue their tiny periodic table in their journal in the center of the page. They can draw the arrows and write their notes all around it, including the definitions above if you like. I have included a page of these for you to copy for this (and other) purposes.

Once that is done, you can go over the arrows, the memory tricks and trends. Then, go through the practice exercises. Finally, if time permits, you can practice any or all of the topics from this unit using the practice sheets.

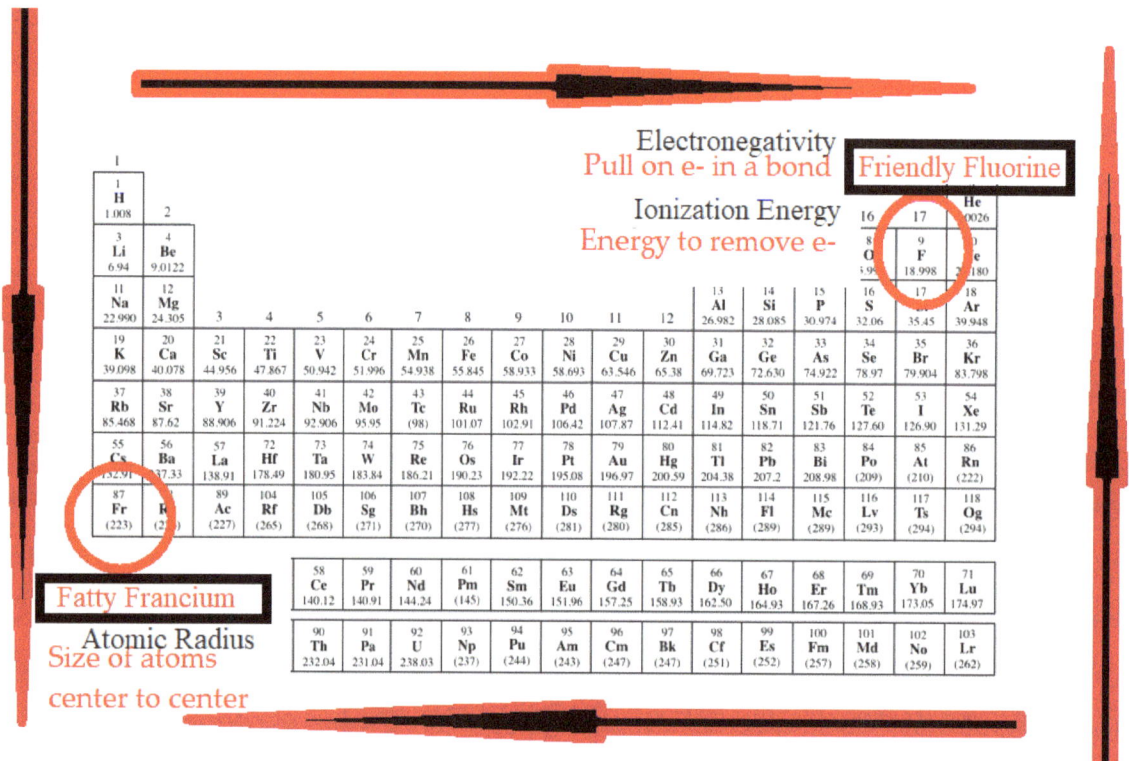

I always start with the blank periodic table and we mark and go from there. I like to start with **'fatty francium'** and I tell them that if that bothers them they can come up with their own (since we live in a PC world, and fatty isn't very nice), but it's true and it works. Fatty means largest, and she is the largest until more elements are discovered and we get a new row on the periodic table. I make sure that I point out that anything to do with 'size' points at her. Whichever element is closer to her is the largest when we compare them, and that's how the trend works.

Draw the arrows to point at her and choose some elements to practice with. I always choose elements that are in the same period (row) and / or family (column), since that is less confusing for learners. Be sure your students know what period and family mean as you do this. Have them pull their PT to the front, putting their fingers on the boxes as you go through and decide. Choose two, such as Potassium and Cesium. Which is larger? Which is smaller? How do you know?

Cesium is larger and Potassium is smaller because Cesium is CLOSER to Fatty Francium. Do as many of these as you like, being aware of your time, since you have more to cover.

When you are ready, you can move on to the two trends that have to do with electrons. **Ionization energy** and **electronegativity** are not the same thing, but they do have to do with how strongly atoms want to pull on or keep electrons.

Electronegativity is for bonding and trying to steal from other elements (if they can). **Ionization energy** is how badly do elements want to keep their valence electrons. Draw the arrows, label **'friendly fluorine'** (I joke that he is really Unfriendly) and tell them how some years my students have named him 'finesse' or 'fake' because he is the big dog on the periodic table – his goal in life is to STEAL electrons from anyone who gets close to him. Talking about this makes it more fun.

Don't worry about too many details, as we will learn much more about this later. For now, we just need to know how to judge by looking at position who is stronger for electronegativity and most likely to give away their electrons for ionization energy.

Choose some elements and again go through how the trend works, such as Silicon and Lead. Which has the higher electronegativity? Silicon, because it is closer to fluorine. Which has the lower ionization energy? Lead, because it is farther away from fluorine. Do a few more if you have time.

Now, for the final piece to this, I want to make sure they understand exactly what low ionization energy means, so I ask, "Who do you think has the lowest ionization energy? Which Family would it be?" That's right, it's the Alkali Metals – they are the farthest group away from fluorine, so they are in the biggest hurry to get rid of their one valence electron (make that connection to yesterday). How about group 2 (the orange ones) – are they in a hurry to get rid of their two? Based on distance from fluorine, yes. As big a hurry as group 1? No, but almost.

I have a selection of you-tube videos that I like to show on the reactivity of group 1 that is perfect for this, but we never have time at this point to show them. If you have longer class periods, I recommend doing that now. Otherwise, I show them later towards the end of the semester to tie this concept with reactivity together.

Magic Board Practice – periodic table concepts. Magic boards are basically a worksheet inside a page protector that we write on with dry erase markers or washable crayons. I have created a set for every unit that I keep in notebooks to pull out as we need them (mine are on colored paper, but you can do white if you prefer), but you could also use the same page protectors over and over if you have the time to trade the pages out as you go instead.

Have the students pick them up as they come in for the day, or hand them out when you finish the lecture. They should spend the rest of class time working on ALL of the concepts from this unit and getting ready for the exam tomorrow.

To check their work, I have a copy of the answers shrunken down on the back of the binder that stores the magic boards. You could also post a copy of them somewhere in your room, but the students need to monitor their own learning.

There are numerous games you can play with these as well, including partner games and table games (groups of 4). They can be placed on a document camera and answers go on white boards to hold up for checking. Use as many styles as you like today and tomorrow to review before the exam – variety makes learning more fun.

Page of Periodic Tables for Notes

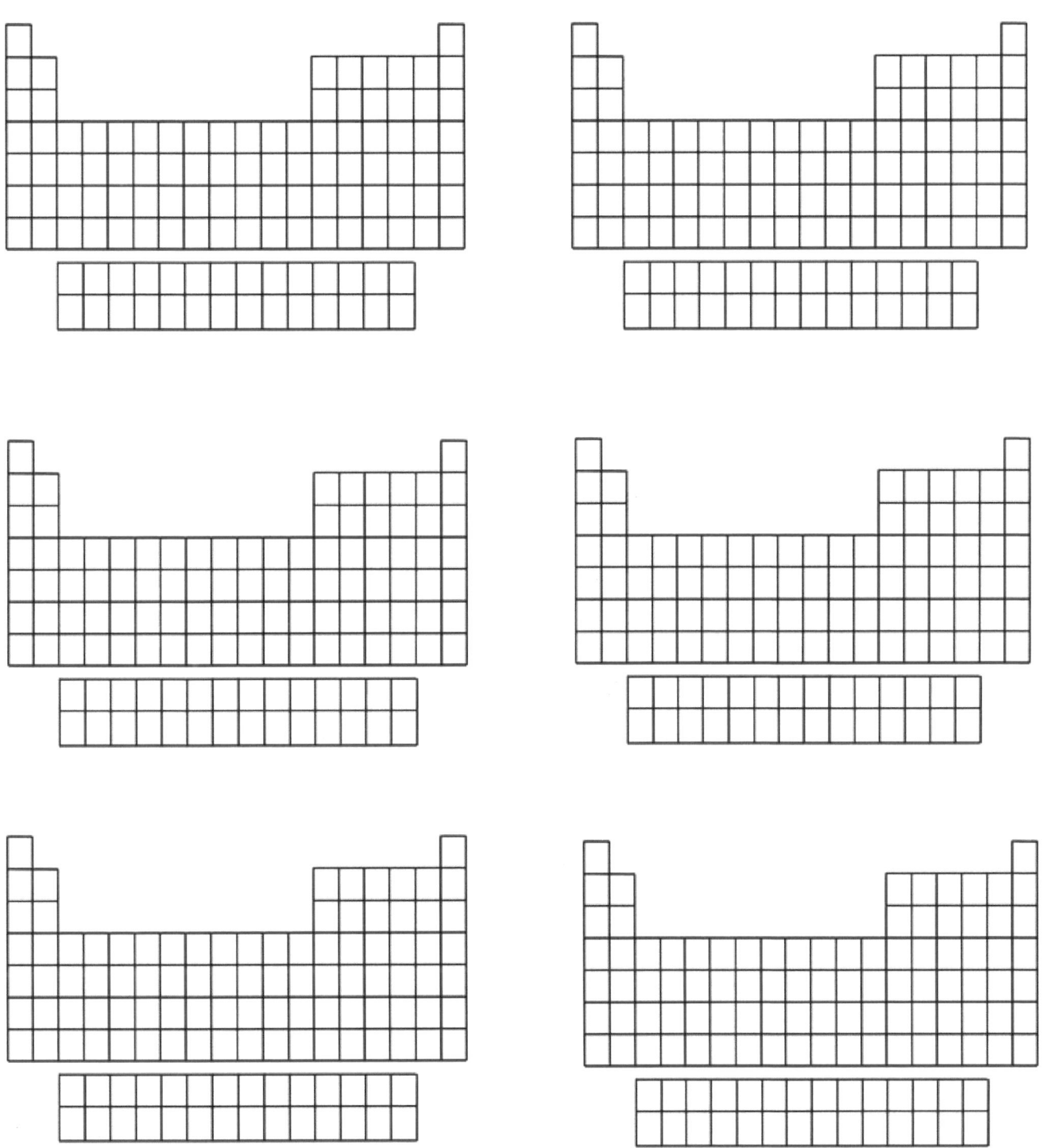

Periodic Table Practice – M&M

Identify who these apply to – **Mendeleev** or **Moseley**

Worked on the periodic table first _____

Revised the periodic table _____

Had to 'fudge' a few of the elements _____

Left blanks for missing elements _____

His version still used today _____

Organized by Atomic Number _____

Organized by Average Atomic Mass _____

© Lavish Publishing, LLC

Periodic Table Practice – M&M ANSWERS

Identify who these apply to – Mendeleev or Moseley

Worked on the periodic table first __**Mendeleev**__

Revised the periodic table __**Moseley**__

Had to 'fudge' a few of the elements __**Mendeleev**__

Left blanks for missing elements **Mendeleev & Moseley**

His version still used today __**Moseley**__

Organized by Atomic Number __**Moseley**__

Organized by Average Atomic Mass __**Mendeleev**__

Periodic Table Practice – Families & Groups

Name the families:　　　　　　　　　　color the sections:

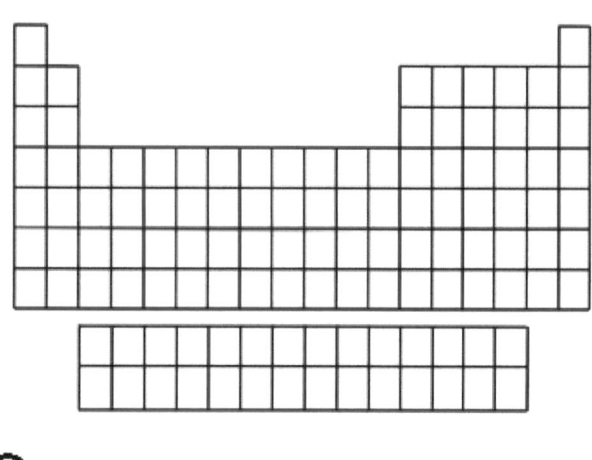
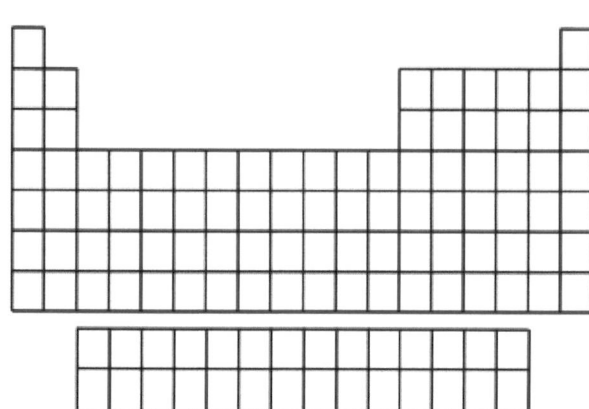

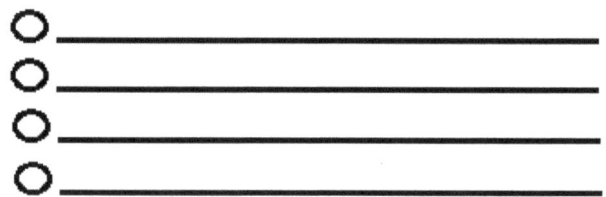

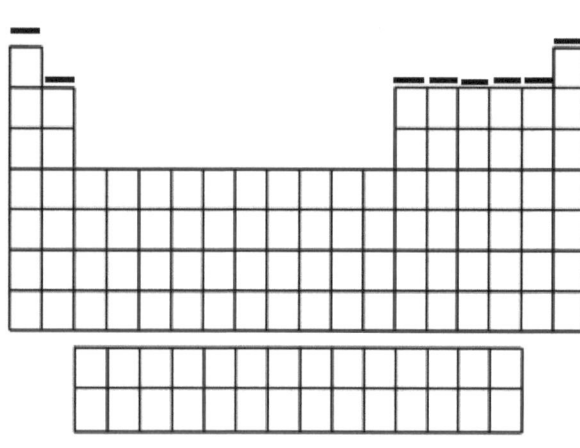
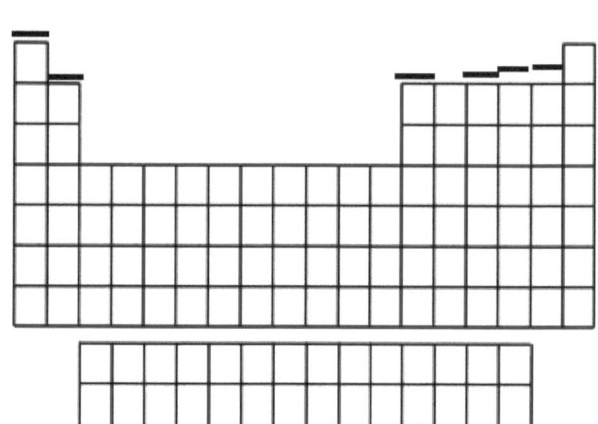

Label the Valence electrons　　　　　Label the ion charges

© Lavish Publishing, LLC

28

Periodic Table Practice – Families & Groups ANSWERS

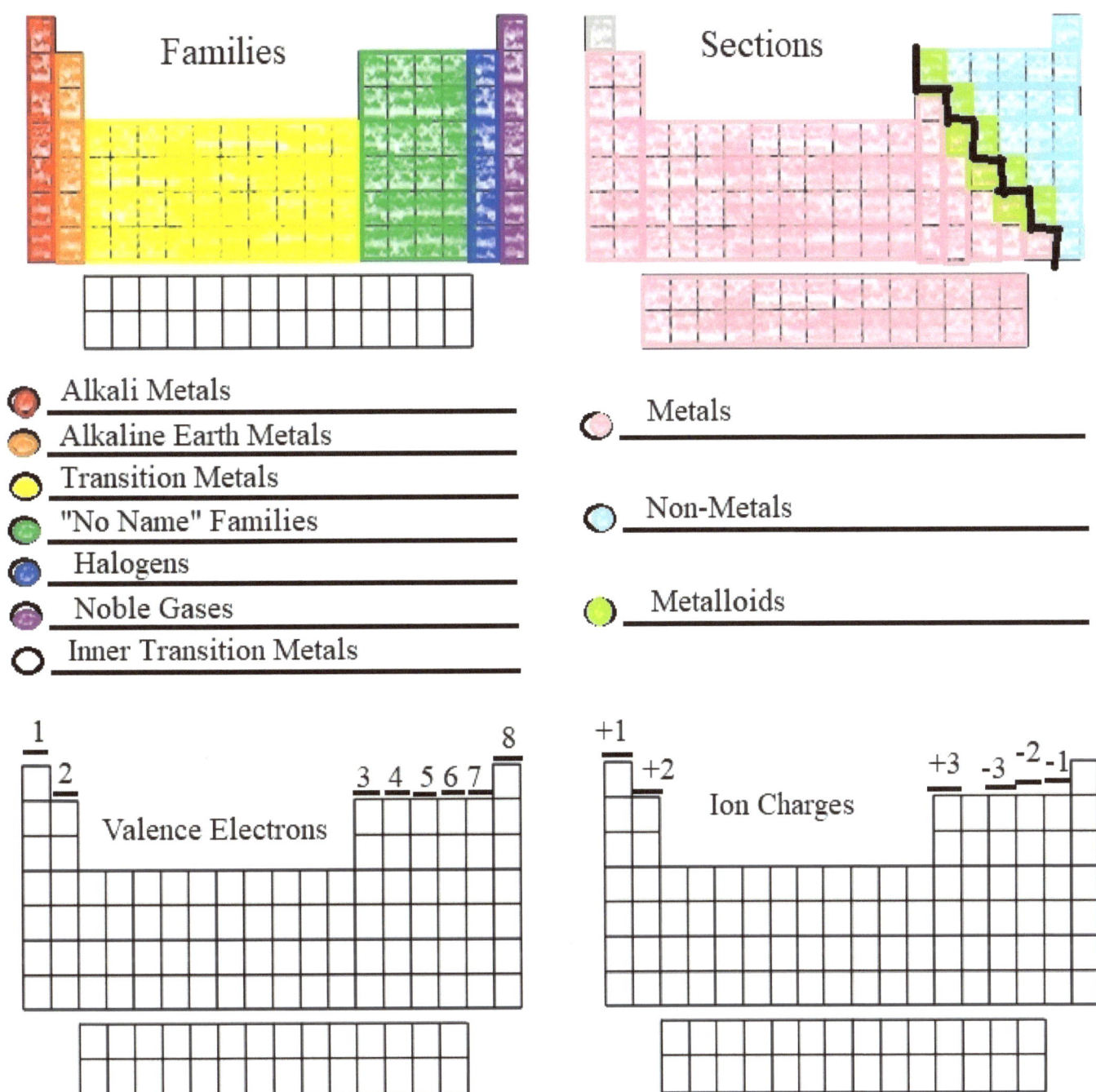

Periodic Table Practice – Table Tricks

Element	Valence e-	Charge	Atom Type (section)
Potassium			
Calcium			
Aluminum			
Boron			
Nitrogen			
Silicon			
Chlorine			
Bromine			
Francium			
Iodine			
Silicon			
Oxygen			

© Lavish Publishing, LLC

Periodic Table Practice – Table Tricks ANSWERS

Element	Valence e-	Ion Charge	Atom Type (section)
Potassium	1	+1	metal
Calcium	2	+2	metal
Aluminum	3	+3	metal
Boron	3	+3	metalloid
Nitrogen	5	-3	Non-metal
Silicon	4	Not an ion	metalloid
Chlorine	7	-1	Non-metal
Bromine	7	-1	Non-metal
Francium	1	+1	metal
Iodine	7	-1	Non-metal
Beryllium	2	+2	metal
Oxygen	6	-2	Non-metal

© Lavish Publishing, LLC

Periodic Table Practice – Trends

Elements	Which has the...	Which has the...
K or Ti	Largest atomic radius?	Highest electronegativity?
Ca or Ba	Lowest ionization energy?	Smallest atomic radius?
Al or S	Lowest electronegativity?	Highest ionization energy?
B or F	Largest atomic radius?	Highest electronegativity?
N or Bi	Lowest ionization energy?	Smallest atomic radius?
Si or S	Lowest electronegativity?	Highest ionization energy?
Cl or Al	Lowest ionization energy?	Smallest atomic radius?
Br or Cu	Lowest electronegativity?	Highest ionization energy?

© Lavish Publishing, LLC

Periodic Table Practice – Trends ANSWERS

Elements	Which has the…	Which has the…
K or Ti	Largest atomic radius? **K**	Highest electronegativity? **Ti**
Ca or Ba	Lowest ionization energy? **Ba**	Smallest atomic radius? **Ca**
Al or S	Lowest electronegativity? **Al**	Highest ionization energy? **S**
B or F	Largest atomic radius? **B**	Highest electronegativity? **F**
N or Bi	Lowest ionization energy? **Bi**	Smallest atomic radius? **N**
Si or S	Lowest electronegativity? **Si**	Highest ionization energy? **S**
Cl or Al	Lowest ionization energy? **Al**	Smallest atomic radius? **Cl**
Br or Cu	Lowest electronegativity? **Cu**	Highest ionization energy? **Br**

© Lavish Publishing, LLC

Topic: Periodic Table Review and Exam
Day: 5
Unit: Periodic Table

Learning Target:
I can demonstrate my understanding of the Periodic Table on a written exam.

Agenda:
A – Red button day
A – Periodic Table Review
A – Periodic Table Exam

Student Materials:
Scantron or answer document
Exam copies

Actions and Rationale:
Red Button Day – as a reminder, classrooms run on routines, and using special testing procedures is one of my most hard fast rules. This is our third exam, but as it is short, we will also do a few small review activities before taking the exam if time allows, so setting up for testing up front is going to help make that possible.

On red button day, ALL of my students' personal stuff goes in a designated location - folding tables at the front of my room.

Their phones go either in their bag or on a charger, which I have two stations in my room – one that is a power strip where they use their charger, and a second set up with my personal chargers that I bought for the class to use – iPhone and android.

For the review activities, you can use the magic boards from day 4 – have them work individually, as partners, or even teams. You can do them as a white board game by asking the questions yourself. The possibilities are endless, so do as much as you have time for until you must begin the exam. It is short and won't take more than twenty minutes once they get started.

After we are seated and ready, I give them my testing rules – starting with the phones. We do the "pocket pat" so they can check to make sure their phones are put away and not accidently back in their pocket, and I tell them what is going to happen if they don't. I like to have them stay seated and working until everyone is finished (I do allow them to slip out to the bathroom QUIETLY). I pick up all the exams at the end, and they are not allowed to get ANY of their stuff until I have all the exams in hand, and I dismiss them to get their stuff.

Phone penalty – if I see them with a phone – using it or not – until they have been dismissed, they get a 1 on the exam (a signal that they have broken testing rules). On the first offense, I let them take the exam again during tutorials the following week and contact their parents to let them know what happened and make the offer of a retake. After that, they get a 1 and it stands. This isn't a game, and they need to learn the consequences of breaking test security.

Yes, I am very strict on this. I am not naive, and I know that kids use their phones to cheat in an unlimited number of ways when given the chance. For this one small piece of my class, I want them to show me what they know on their own, and I don't want them sharing copies of my test in any way. If you are strict and diligent, the number of problems will be far less in the long run. Be clear and up front with your expectations and stick to your rules. The first few exams they may whine and complain, but it should die down. Anyone who is over the top or causes disruption of the test over it gets a phone call home so I can chat with their parents about why their student can't follow the rules like everyone else.

For the **answer documents**, we use scantrons, which makes grading super easy, but if you don't have access, you can always have them fill out their answers on a strip of paper, which will make them easier to grade. I also like to provide each student a copy of the exam so they can write on them, which I always print and shrink to fit on a single page. I like to do this so they can practice using test taking strategies as they work. I also keep them locked up before the test. After they are used, I keep them locked in another cabinet for the rest of the year in case we need to go back and look at someone's test, so I do ask them to put their names on them.

How ever you go, make sure that you get ALL of your tests and answer documents back before students get their stuff or leave the room.

M & M Quiz

1. Which worked on the Periodic Table first?

2. What did Mendeleev use to organize his periodic table?

3. Why was Moseley's version better?

© Lavish Publishing, LLC

M & M Quiz

1. Which worked on the Periodic Table first?

2. What did Mendeleev use to organize his periodic table?

3. Why was Moseley's version better?

© Lavish Publishing, LLC

M & M Quiz

1. Which worked on the Periodic Table first?

2. What did Mendeleev use to organize his periodic table?

3. Why was Moseley's version better?

© Lavish Publishing, LLC

M & M Quiz

1. Which worked on the Periodic Table first?

2. What did Mendeleev use to organize his periodic table?

3. Why was Moseley's version better?

© Lavish Publishing, LLC

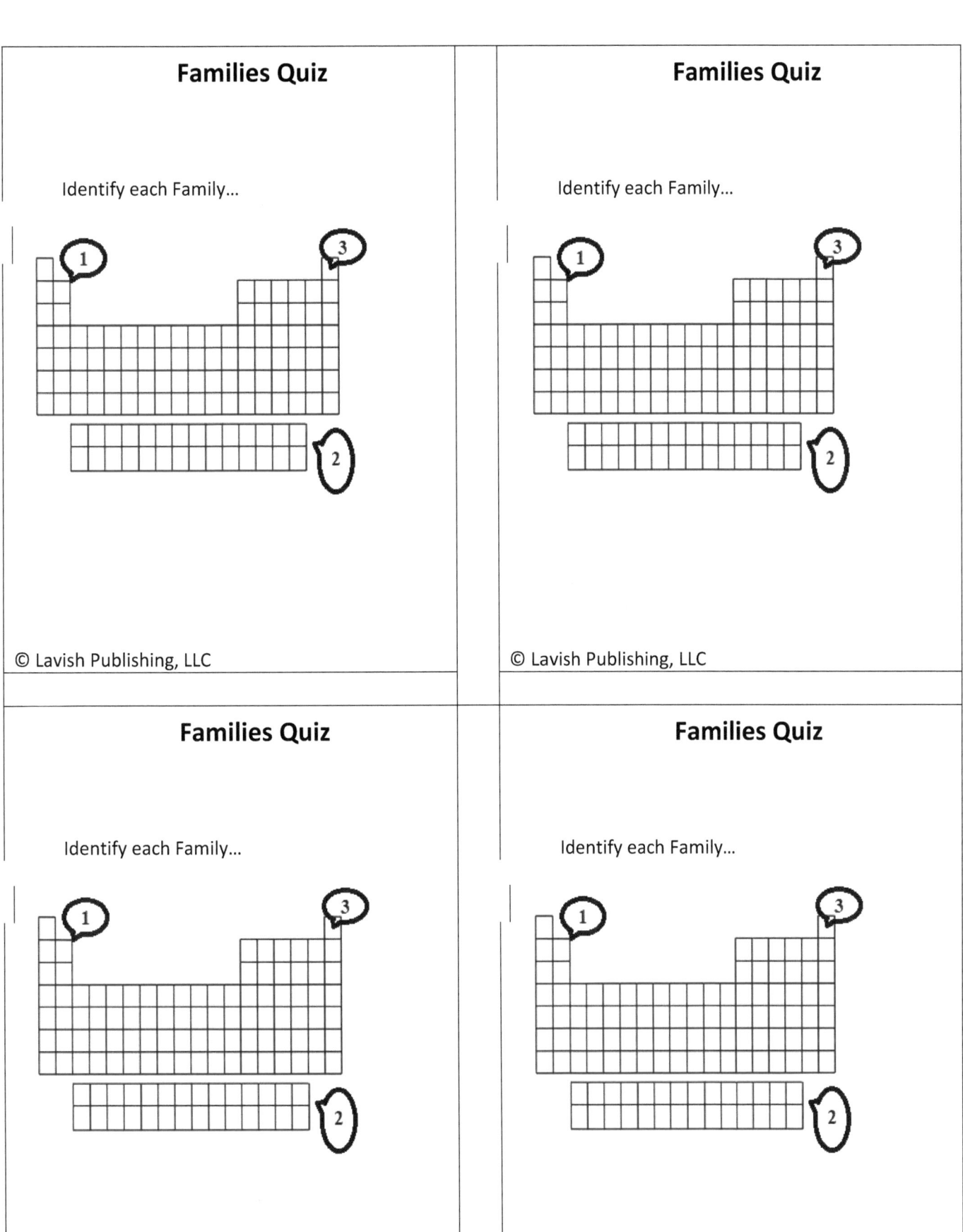

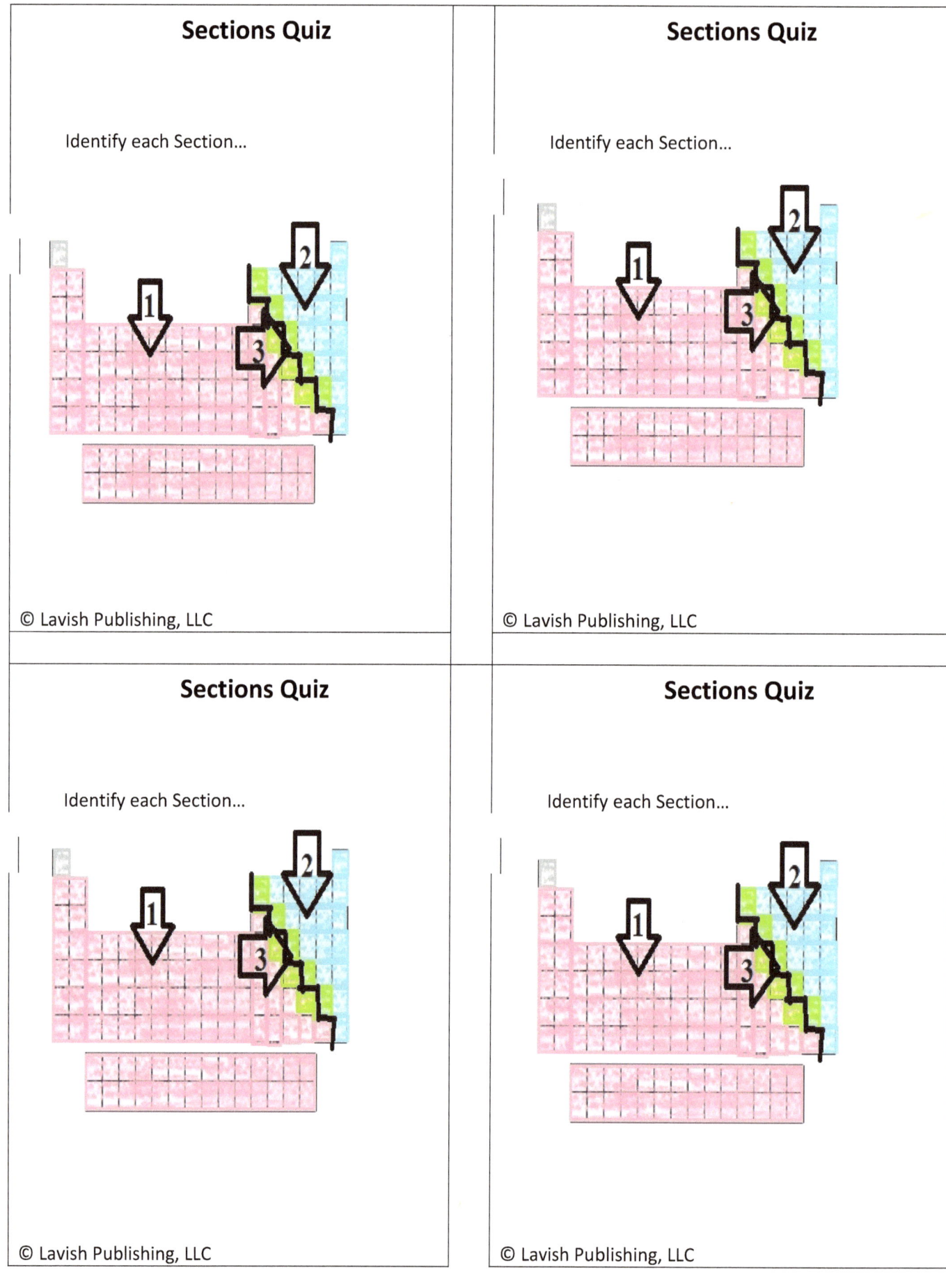

Table Tricks Quiz

Given Carbon, Hydrogen, Nitrogen and Oxygen...

1. Which has 1 valence electron?

2. Which has a -3 Charge?

3. Which one doesn't have a charge?

© Lavish Publishing, LLC

Table Tricks Quiz

Given Carbon, Hydrogen, Nitrogen and Oxygen...

1. Which has 1 valence electron?

2. Which has a -3 Charge?

3. Which one doesn't have a charge?

© Lavish Publishing, LLC

Table Tricks Quiz

Given Carbon, Hydrogen, Nitrogen and Oxygen...

1. Which has 1 valence electron?

2. Which has a -3 Charge?

3. Which one doesn't have a charge?

© Lavish Publishing, LLC

Table Tricks Quiz

Given Carbon, Hydrogen, Nitrogen and Oxygen...

1. Which has 1 valence electron?

2. Which has a -3 Charge?

3. Which one doesn't have a charge?

© Lavish Publishing, LLC

Trends Quiz

Given Carbon, Silicon, Sulfur and Oxygen…

1. Which has the largest atomic radius?

2. Which has the lowest ionization energy?

3. Which one has the highest electronegativity?

© Lavish Publishing, LLC

Trends Quiz

Given Carbon, Silicon, Sulfur and Oxygen…

1. Which has the largest atomic radius?

2. Which has the lowest ionization energy?

3. Which one has the highest electronegativity?

© Lavish Publishing, LLC

Trends Quiz

Given Carbon, Silicon, Sulfur and Oxygen…

1. Which has the largest atomic radius?

2. Which has the lowest ionization energy?

3. Which one has the highest electronegativity?

© Lavish Publishing, LLC

Trends Quiz

Given Carbon, Silicon, Sulfur and Oxygen…

1. Which has the largest atomic radius?

2. Which has the lowest ionization energy?

3. Which one has the highest electronegativity?

© Lavish Publishing, LLC

Form A Periodic Table Vocabulary Quiz
© Lavish Publishing, LLC

1		___ are found along the stairs (right above or below) on the periodic table - they take on a variety of characteristics from both metals and non-metals.
2		___ are found on top of the stairs on the periodic table - they are dull, brittle solids, can be gases, have a negative charge in ion form (takers) and are poor conductors of heat and electricity (insulators).
3		The 8th group or family are known as the ___ - because they satisfy the 'octet rule' alone, they do not "hob-knob" with other elements or form bonds.
4		___ is equal to protons plus neutrons; it is the average atomic mass rounded to a whole number.
5		___ is the process of deriving the strict logical consequences of assumed premises (what do you think happened / will happened based on the data).
	word choices	(A) Inference (B) Mass Number (C) Metalloids (D) Noble Gases (E) Non-metals

Form B Periodic Table Vocabulary Quiz
© Lavish Publishing, LLC

1		____ is equal to the number of protons in an atom; used to arrange the periodic table.
2		____ is the tendency for an atom to attract electrons to itself when it is chemically combined with another element.
3		A Russian born scientist named Dmitri ___ created the first periodic table of elements based upon the atomic mass of the elements - he was able to use this to make predictions about the behavior of elements.
4		Columns on the periodic table are called ____.
5		Elements in the same group or column on the periodic table are said to belong to the same ___.
	word choices	(A) Atomic Number (B) Chemical Family (C) Electronegativity (D) Groups (E) Mendeleev

Form C **Periodic Table Vocabulary Quiz**
© Lavish Publishing, LLC

1		Group I of the Periodic Table, the ____, is composed of highly reactive metals.
2		Henry ___ was able to take Mendeleev's periodic table and rearrange it in order by atomic number - which made it work better - and is still used today.
3		On the periodic table, we can see that there are several ___ that repeat, period after period.
4		The ___ are found in the center of the periodic table, and get their name because they have a tendency to change their oxidation numbers by moving their valence electrons around.
5		The ___ is the energy required to completely remove an electron from a gaseous atom or ion.
	word choices	(A) Alkali Metals (B) Ionization Energy (C) Moseley (D) transition metals (E) trends

Form D **Periodic Table Vocabulary Quiz**
© Lavish Publishing, LLC

1		The ___ is the size of the radius of an ion.
2		The ___ of an element is half of the distance between the centers of two atoms of that element that are just touching each other.
3		The ___, the 2nd group on the periodic table, are not found freely in nature.
4		___ are found UNDER the stairs on the periodic table - they are shiny, hard, have a + charge in ion form (givers), and are good conductors of heat and electricity.
5		The elements of Group VII of the Periodic Table are called ___, which means "salt formers".
	word choices	(A) Alkaline Earth Metals (B) Atomic Radius (C) halogens (D) metals (E) valence

Periodic Table Exam

Choose the best answer for each question.

1. Which family is Sodium (Na) a part of?
A. Alkali Metals
B. Alkaline Earth Metals
C. Halogens
D. Noble Gases

2. Which family is Fluorine (F) a part of?
A. Alkali Metals
B. Alkaline Earth Metals
C. Halogens
D. Noble Gases

3. Based on position, which has the largest atomic radius?
A. Potassium (K)
B. Cesium (Cs)
C. Cobalt (Co)
D. Iridium (Ir)

4. Based on position, which has the largest electronegativity?
A. Potassium (K)
B. Cesium (Cs)
C. Cobalt (Co)
D. Iridium (Ir)

5. How did Moseley organize his version of the Period Table?
A. average atomic mass
B. simple atomic mass
C. atomic symbol
D. atomic number

6. Which group of elements is found above the stairs?
A. metals
B. non-metals
C. metalloids
D. heavy metals

7. Which group of elements will have a positive charge?
A. metals
B. non-metals
C. metalloids
D. heavy metals

8. Which group of elements will generally take or steal electrons?
A. metals
B. non-metals
C. metalloids
D. heavy metals

9. Which group of elements will have electrical conductivity?
A. metals
B. non-metals
C. metalloids
D. heavy metals

© Lavish Publishing, LLC

10. Which group of elements will have properties that are hard to predict?
A. metals
B. non-metals
C. metalloids
D. heavy metals

11. How many valence electrons does oxygen have?
A. 2
B. 6
C. 8
D. 16

12. How many valence electrons does Krypton have?
A. 2
B. 6
C. 8
D. 18

13. Which rule tells us that atoms want to have eight valence electrons so they have full valence shells like the noble gases?
A. Periodic Law
B. Octet Rule
C. Moseley's Rule
D. there is no such rule

14. Which group of elements cannot have 'table tricks' applied to it?
A. metals
B. non-metals
C. Transition Metals
D. Noble Gases

15. Which trend has to do with an atom's ability to STEAL or TAKE electrons from other atoms?
A. electronegativity
B. ionization energy
C. atomic radius
D. ionic radius

16. Which trend has to do with an atom's ability to LOSE or GIVE AWAY electrons to other atoms?
A. electronegativity
B. ionization energy
C. atomic radius
D. ionic radius

17. Which trend has to do with the size of a pair of atoms side by side?
A. electronegativity
B. ionization energy
C. ionic radius
D. atomic radius

© Lavish Publishing, LLC

Quizzes Answer Keys

M & M Quiz

1. Which worked on the Periodic Table first?
   Mendeleev

2. What did Mendeleev use to organize his periodic table?
   Ave Atomic Mass

3. Why was Moseley's version better?
   The periods created gave the new periodic table predictive powers

© Lavish Publishing, LLC

Families Quiz

Identify each Family...

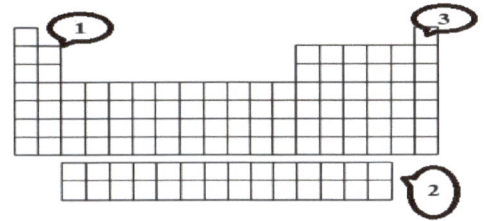

1. **Alkaline Earth Metals**
2. **Inner Transition Metals**
3. **Noble Gases**

© Lavish Publishing, LLC

Sections Quiz

Identify each Section...

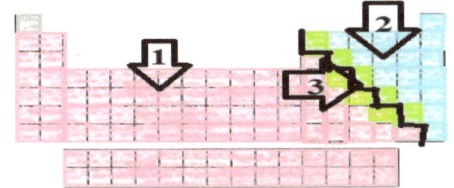

1. **Metals**
2. **Non-Metals**
3. **Metalloids**

© Lavish Publishing, LLC

Table Tricks Quiz

Given Carbon, Hydrogen, Nitrogen and Oxygen...

1. Which has 1 valence electron?
   hydrogen

2. Which has a -3 Charge?
   nitrogen

3. Which one doesn't have a charge?
   carbon

© Lavish Publishing, LLC

Trends Quiz

Given Carbon, Silicon, Sulfur and Oxygen...

1. Which has the largest atomic radius?

   Silicon

2. Which has the lowest ionization energy?

   Silicon

3. Which one has the highest electronegativity?

   Oxygen

© Lavish Publishing, LLC

Periodic Table Vocabulary Quiz Answers

Form A		Form B		Form C		Form D	
1	(C) metalloids	1	(A) atomic number	1	(A) alkali metals	1	(E) valence
2	(E) non-metals	2	(C) electronegativity	2	(C) Moseley	2	(B) atomic radius
3	(D) noble gases	3	(E) Mendeleev	3	(E) trends	3	(A) alkaline earth metals
4B	(B) mass number	4	(D) groups	4	(D) transition metals	4	(D) metals
5	(A) inference	5	(B) chemical family	5	(B) ionization energy	5	(C) halogens

Periodic Table Exam Answers

1. A
2. C
3. B
4. C
5. D
6. A
7. A
8. B
9. A
10. C
11. B
12. C
13. B
14. C
15. A
16. B
17. D

About the Author

Born and raised in West Texas, Sammie Jacobs aspired to be a teacher at an early age but did not achieve her dream until the age of 38 when she earned her composite science certification in 2008.

Answering the call of the classroom, she went to work at a local high school teaching Chemistry. Through all the years, she loved the subject and her students. Combining her dedication to both, Sam continuously searched for ways to improve her instruction and meet her students varied needs, leading her to create and construct many of her materials.

Mentoring new teachers as they came into the district and her department, Sam could see the importance of helping those new to her beloved field. First, she began to build the files that would become the foundation of this series. Later, she realized that putting those lesson plans and tools into the hands of others could mean a great deal to her colleagues and countless students, eventually even those across the country or around the world.

Always striving for more and looking for the best in herself and those around her, Sammie Jacobs is releasing this complete version of her Chemistry lessons – 17 units in all, which will be available for purchase at a nominal fee in paperback format. These files contain everything needed to plan and execute a solid foundational year of Chemistry for any teacher, be it veteran or novice.

But Sam also knows there is more that can be done, so she is also founding a teacher community group on Facebook. Open to everyone who works as a science educator, this is Sam's legacy, her dream coming true. If you are a veteran with advice to share or a novice looking for a helpful hand, come and join and let us grow together. In the end, Sam hopes many will find these tools useful and many more will be inspired to reach for their dreams, whatever they might be…

Units in the Teaching Chemistry in a Diversified Classroom Series

Lab Safety & Equipment
Properties of Matter
Periodic Table
Atomic Structure
Electrons in Atoms
Ionic Bonding
Covalent Bonding
Names & Formulas
Equations & Reactions
Calculations
Stoichiometry
States of Matter
Gases
Solutions
Acids & Bases
Thermochemistry
Nuclear Chemistry

Follow the Teaching Chemistry Series on Facebook –
https://www.facebook.com/SammieJacobsChemistry/?

Join the Chemistry Professional Learning Community on Facebook –
https://www.facebook.com/groups/TeachingChemistryPLC/